PLANET PARADE

*A COMPREHENSIVE GUIDE TO OBSERVING,
AND UNDERSTANDING THE ASTRONOMICAL
PHENOMENON OF PLANETARY ALIGNMENT*

HARPER PIPER

Copyright HARPER PIPER, 2024.

All rights reserved. No part of this publication may be reproduced, distributed, or transmitted in any form or by any means, including photocopying, recording, or other electronic or mechanical methods, without the prior written permission of the publisher, except in the case of brief quotations embodied in critical reviews and certain other non-commercial uses permitted by copyright law.

Table of Contents

Introduction _____ 7
The Fascination with the Night Sky _____ 7

Chapter 1 _____ 11
What is a Planetary Alignment? _____ 11
Definition and Explanation _____ 11
Historical Significance and Observations _____ 14
Astronomical vs. Visual Alignments _____ 17

Chapter 2 _____ 23
The Science Behind Planetary Movements _____ 23
Orbital Mechanics of the Solar System _____ 23
The Ecliptic Plane and Its Significance _____ 29
How Alignments Occur _____ 32
Alignment Types and Conditions _____ 33
Mathematical and Geometrical Considerations _____ 35
Predicting Alignments _____ 36

Chapter 3 _____ 39
The June 3, 2024 Planetary Parade _____ 39
Overview of the Event _____ 39
Profiles of the Six Planets Involved _____ 40
Viewing Tips and Equipment _____ 45

Chapter 4 — 49

Observing the Alignment — 49
- Best Times and Locations — 49
- Necessary Equipment: Binoculars, Telescopes, Apps — 51
- Challenges and Tips for Observation — 53

Chapter 5 — 55

The Role of Each Planet in the Alignment — 55
- Mercury: The Elusive Inner Planet — 55
- Mars: The Red Planet — 56
- Jupiter: The Giant Planet — 57
- Saturn: The Ringed Planet — 57
- Uranus and Neptune: The Distant Ice Giants — 58

Chapter 6 — 59

Future Planetary Alignments — 59
- Upcoming Alignments in 2024 and 2025 — 59
- Rare Multi-Planet Alignments in the Future — 60
- Long-term Predictions and Historical Patterns — 61

Chapter 7 — 63

Historical and Cultural Perspectives — 63
- Ancient Observations and Myths — 63
- Cultural Significance Across Civilizations — 64
- Modern-Day Interest and Media Coverage — 66

Chapter 8 — 67

The Impact of Planetary Alignments on Earth — 67
- Debunking Myths and Misconceptions — 67
- The Influence on Tides, Seasons, and Astrology — 68
- Scientific Studies and Findings — 69

Chapter 9 **_____ *71***
   Tools for Stargazers _____ 71
      Apps and Software for Tracking Celestial Events _____ 71
      Using Star Maps and Charts _____ 73

Introduction

The Fascination with the Night Sky

Since time immemorial, the night sky has captivated human imagination. Ancient civilizations gazed upon the stars, weaving stories and myths around the celestial bodies they observed. For them, the night sky was not just a vast expanse of darkness punctuated by twinkling lights but a cosmic canvas teeming with gods, heroes, and creatures. These early stargazers were the pioneers of astronomy, laying the foundations for a field of study that continues to intrigue us to this day.

The allure of the night sky lies in its mystery and grandeur. As the sun sets and the veil of night descends, a new world emerges, one that is both familiar and alien. The constellations, planets, and the Milky Way galaxy appear, inviting us to ponder our place in the universe. The night sky evokes a sense of wonder and curiosity, prompting us to ask profound questions: What are these lights? How far away are they? Are we alone in the universe?

This fascination has led to significant advancements in our understanding of the cosmos. Ancient astronomers like Ptolemy and Copernicus, as well as modern scientists like Galileo and Hubble, have contributed to our knowledge of the universe, each building on the work of their predecessors. Their discoveries have revealed the night sky to be a dynamic and ever-changing realm, governed by the laws of physics and mathematics.

But the night sky is not just a subject for scientific inquiry; it has also inspired art, literature, and philosophy. From Van Gogh's "Starry Night" to the poetry of John Keats, the beauty of the cosmos has been a source of inspiration and reflection. The stars have been used for navigation, timekeeping, and even as a means of predicting the future through astrology. They have guided sailors across oceans, informed agricultural practices, and shaped religious and cultural beliefs.

One of the most mesmerizing phenomena that occur in the night sky is the alignment of planets. Unlike the fixed stars, planets move relative to each other and to the backdrop of the zodiac. When they come together in alignment, it creates a spectacle that captures the imagination of both astronomers and casual observers. These alignments, also known as planetary parades, have been observed and recorded throughout history, each one offering a unique opportunity to witness the intricate dance of the solar system.

Understanding planetary alignments requires a grasp of both astronomy and history. By studying these events, we can learn about the movements and interactions of celestial bodies, as well as the ways in which humans have interpreted and responded to them. This book aims to provide a comprehensive exploration of planetary alignments, blending scientific explanation with historical and cultural context.

By following this structure, the book aims to be both informative and engaging, providing a comprehensive overview of planetary alignments while also offering practical advice and personal insights. Whether you are a seasoned astronomer or a curious newcomer, this book will guide you through the fascinating world of planetary alignments and help you make the most of the upcoming event on June 3, 2024.

Chapter 1

What is a Planetary Alignment?

Definition and Explanation

A planetary alignment occurs when several planets in our solar system appear to line up in the sky from our perspective on Earth. This celestial phenomenon can involve two or more planets appearing close together, forming a line or nearly a line in the night sky. It's important to note that while the planets may seem aligned from our viewpoint on Earth, they are not necessarily in a straight line in space. Instead, their positions in their respective orbits create the illusion of alignment.

To understand planetary alignments, it is essential to grasp the basic mechanics of the solar system. Our solar system consists of the Sun and the celestial bodies that orbit it, including eight major planets, their moons, dwarf planets, and various small solar system bodies such as asteroids and comets. Each planet follows an elliptical orbit around the Sun, with the orbits lying roughly within the same plane, known as the ecliptic plane.

Planetary alignments are a result of the relative positions and motions of the planets along their orbits. Each planet moves at a different speed depending on its distance from the Sun, as described by Kepler's laws of planetary motion. These laws explain that planets move in elliptical orbits with the Sun at one focus, they sweep out equal areas in equal times, and the square of the orbital period is proportional to the cube of the semi-major axis of the orbit. These principles help us predict when and where alignments will occur.

There are different types of planetary alignments, including **conjunctions, oppositions, and syzygies.** A *conjunction* occurs when two planets appear very close to each other in the sky, often within a few degrees of separation. For example, a conjunction of Venus and Jupiter might make them appear as a bright pair in the evening sky. **Oppositions** occur when planets are on opposite sides of the Earth, making them appear on opposite sides of the sky. This typically applies to outer planets like Mars, Jupiter, and Saturn. *Syzygy* is a more general term that refers to any alignment of three or more celestial bodies.

Alignments can also involve other celestial bodies such as the Moon or the Sun. For instance, during a solar eclipse, the Moon aligns between the Earth and the Sun, creating a syzygy. Similarly, a lunar eclipse occurs when the Earth aligns between the Sun and the Moon. These events are specific types of syzygies involving the Earth, Moon, and Sun.

Planetary alignments vary in their rarity and significance. Some alignments are relatively common, occurring every few years, while others are extremely rare, happening only once in several centuries. The degree of alignment, or how close the planets appear to be to one another, also varies. Some alignments are so close that the planets seem to merge into a single bright point, while others involve a looser grouping.

In addition to visual alignment, where the planets appear close together in the sky, there is also true alignment, where the planets are positioned in a straight line in three-dimensional space. True alignments are much rarer and often involve complex gravitational interactions between the planets.

Historical Significance and Observations

One of the earliest recorded observations of planetary alignments comes from the Babylonians, who were adept astronomers. They developed sophisticated mathematical methods to predict the movements of celestial bodies and recorded planetary alignments on clay tablets. The Babylonians believed that the positions and movements of the planets influenced the affairs of kings and empires, and they used these alignments for astrological predictions.

In ancient Greece, philosophers and astronomers like Pythagoras and Ptolemy studied planetary alignments as part of their exploration of the cosmos. Ptolemy's geocentric model of the universe, which placed the Earth at the center with planets orbiting in complex paths called epicycles, attempted to explain the observed motions of the planets, including alignments. Although this model was later replaced by the heliocentric model proposed by Copernicus, it was a crucial step in the development of astronomical theories.

The significance of planetary alignments can also be seen in ancient Chinese astronomy. Chinese astronomers meticulously recorded celestial events, including planetary alignments, in their official records. These observations were believed to be omens that could foretell natural disasters, the rise and fall of dynasties, and other significant events. The Chinese developed a complex system of astrology and divination based on these celestial phenomena.

In medieval Europe, planetary alignments continued to be studied and interpreted through the lens of astrology. Astrologers believed that the positions of the planets at the time of a person's birth influenced their personality and destiny. Planetary alignments were thought to have a profound impact on human affairs, and astrological charts were used to guide decisions ranging from personal relationships to political strategies.

One of the most famous examples of a planetary alignment's historical significance is the Great Conjunction of 7 BCE, which is often linked to the

Star of Bethlehem mentioned in the Christian nativity story. This conjunction of Jupiter and Saturn was a rare and spectacular event that could have been interpreted as a significant astronomical sign by the Magi, leading them to Bethlehem.

In more recent history, the scientific revolution brought a deeper understanding of planetary alignments through the work of astronomers like Johannes Kepler and Isaac Newton. Kepler's laws of planetary motion, formulated in the early 17th century, provided a mathematical framework for predicting planetary positions and alignments. Newton's laws of motion and universal gravitation further explained the forces governing planetary movements, laying the groundwork for modern celestial mechanics.

The invention of the telescope in the early 17th century revolutionized the observation of planetary alignments. Galileo Galilei's telescopic observations of Jupiter's moons and the phases of Venus provided crucial evidence for the heliocentric model of the solar system, challenging the geocentric view that had dominated for centuries. Telescopes allowed astronomers to observe planetary alignments with greater precision and detail, leading to significant advancements in our understanding of the cosmos.

Astronomical vs. Visual Alignments

It is crucial to distinguish between astronomical and visual alignments when discussing planetary alignments. While both types involve the apparent positioning of planets in the sky, they differ in terms of their underlying principles and observational significance.

Astronomical alignments refer to the true positional relationships between planets in three-dimensional space. These alignments are governed by the precise orbital mechanics of the solar system and involve the actual spatial arrangement of planets along their orbits. For an astronomical alignment to occur, the planets must be positioned in such a way that they form a straight line, or nearly a straight line, in space. This type of alignment is much rarer than visual alignments due to the vast distances and varying orbital planes of the planets.

True astronomical alignments have significant implications for gravitational interactions between planets. When multiple planets align in space, their combined gravitational forces can have measurable effects on their orbits. These gravitational interactions can lead to perturbations in the orbits, causing slight deviations over time. While these effects are generally small and do not pose any immediate danger to Earth, they are of great interest to astronomers studying the dynamics of the solar system.

One famous example of an astronomical alignment is the so-called **"Grand Alignment"** that occurred in 2000, when the planets Mercury, Venus, Earth, Mars, Jupiter, and Saturn were roughly aligned on the same side of the Sun. Although not perfectly aligned in a straight line, this configuration was significant enough to attract widespread attention. The gravitational effects of this alignment were carefully studied by astronomers to understand their impact on the solar system.

In contrast, visual alignments refer to the apparent positioning of planets as seen from a specific vantage point, usually Earth. These alignments are influenced by our line of sight and can create the illusion that planets are close together in the sky. Visual alignments do not necessarily correspond to true spatial alignments but are instead a result of the planets' positions in their orbits relative to Earth.

Visual alignments are more common than true astronomical alignments and can be observed without the need for sophisticated equipment. They are primarily of interest to amateur astronomers, stargazers, and the general public due to their aesthetic appeal and ease of observation. These alignments provide opportunities to observe multiple planets in a single view, often making for striking celestial displays.

A well-known example of a visual alignment is the Great Conjunction of 2020, when Jupiter and Saturn appeared closer together in the sky than they had been since 1623. Although Jupiter and Saturn were not truly aligned in space, their positions relative to Earth made them appear almost as a single bright point. This event was visible to the naked eye and generated considerable excitement, with people around the world turning their gaze to the night sky to witness the spectacle.

Visual alignments can involve various combinations of planets and can occur in different parts of the sky at different times of the year. Some visual alignments, such as those involving the inner planets Mercury and Venus, can be observed just after sunset or just before sunrise. Others, involving outer planets like Mars, Jupiter, and Saturn, can be observed throughout the night.

In addition to planetary alignments, visual alignments can also include alignments with other celestial objects such as the Moon and bright stars. For example, a conjunction of Venus and the Moon can create a stunning sight, with the bright planet appearing close to the crescent Moon. Such events are often eagerly anticipated by stargazers and are well-documented in astronomical calendars.

The distinction between astronomical and visual alignments highlights the importance of perspective in celestial observations. While true astronomical alignments are rare and scientifically significant, visual alignments offer more frequent opportunities for observation and enjoyment. Both types of alignments contribute to our understanding of the dynamics of the solar system and our appreciation of the beauty of the night sky.

Chapter 2

The Science Behind Planetary Movements

Orbital Mechanics of the Solar System

The motion of planets in our solar system is governed by a set of principles and laws that have been developed and refined over centuries. The foundation of our understanding of these movements lies in the laws of orbital mechanics, which describe the motion of celestial bodies under the influence of gravity. The primary contributors to these principles are Johannes Kepler and Sir Isaac Newton.

Kepler's Laws of Planetary Motion

Johannes Kepler, a German mathematician and astronomer, formulated three fundamental laws of planetary motion in the early 17th century. These laws describe the orbits of planets around the Sun and are essential for understanding how planetary alignments occur.

1. Kepler's First Law (Law of Ellipses):

- **Statement:** Every planet orbits the Sun in an elliptical path, with the Sun at one of the two foci of the ellipse.

- **Explanation:** Unlike a perfect circle, an ellipse is an elongated shape with two focal points. For a planet orbiting the Sun, the Sun occupies one of these foci. This means that the distance between a planet and the Sun varies as the planet moves along its orbit. When a planet is closest to the Sun (perihelion), it moves faster; when it is farthest from the Sun (aphelion), it moves slower.

2. Kepler's Second Law (Law of Equal Areas):

- **Statement:** A line segment joining a planet and the Sun sweeps out equal areas during equal intervals of time.

- **Explanation:** This law implies that a planet's speed varies depending on its distance from the Sun. When a planet is near perihelion, it travels more rapidly, covering a larger arc in a shorter time.

Conversely, near aphelion, the planet moves more slowly, sweeping out a smaller arc in the same duration. This ensures that the area swept by the line segment remains constant over equal time periods.

3. Kepler's Third Law (Law of Harmonies):

- **Statement:** The square of a planet's orbital period (the time it takes to complete one orbit around the Sun) is proportional to the cube of the semi-major axis of its orbit.

- **Mathematical Formulation:** $(T^2 \propto a^3)$

- **Explanation:** The orbital period (T) is the time taken for a planet to complete one full orbit around the Sun, and the semi-major axis (a) is the average distance from the planet to the Sun. This law indicates that planets farther from the Sun have longer orbital periods, moving more slowly along their larger orbits.

Newton's Laws of Motion and Universal Gravitation

Isaac Newton built upon Kepler's work with his own groundbreaking contributions to physics and astronomy. Newton's laws of motion and his law of universal gravitation provide a comprehensive framework for understanding the forces that govern planetary movements.

1. Newton's First Law of Motion (Law of Inertia):

- **Statement:** An object at rest stays at rest, and an object in motion continues in motion with the same speed and in the same direction unless acted upon by an external force.

- **Implication:** In the context of planetary motion, this law means that a planet will continue moving in its orbital path unless an external force, such as gravitational pull from another celestial body, acts upon it.

2. Newton's Second Law of Motion (Law of Acceleration):

- **Statement:** The acceleration of an object is directly proportional to the net force acting upon it and inversely proportional to its mass.

- **Mathematical Formulation:** ($F = ma$)

- **Implication:** This law explains how the gravitational force exerted by the Sun affects a planet's motion. The force causes the planet to accelerate towards the Sun, maintaining its orbit.

3. Newton's Third Law of Motion (Action and Reaction):

- **Statement:** For every action, there is an equal and opposite reaction.

- **Implication:** This law is evident in the mutual gravitational attraction between the Sun and the planets. While the Sun exerts a gravitational force on the planets, the planets also exert an equal and opposite force on the Sun. However, due to the Sun's much larger mass, its motion is negligible compared to that of the planets.

4. Newton's Law of Universal Gravitation:

- **Statement:** Every mass exerts an attractive force on every other mass. The magnitude of this force is directly proportional to the product of the two masses and inversely proportional to the square of the distance between them.

- **Mathematical Formulation:** ($F = G m_1 m_2 / r^2$)

- **Explanation:** In this equation, (F) is the gravitational force between two masses (m_1) and (m_2), (G) is the gravitational constant, and (r) is the distance between the centers of the two masses. This law explains the gravitational pull that the Sun exerts on the planets, keeping them in their orbits, and also the gravitational interactions between planets that can influence their paths.

The Ecliptic Plane and Its Significance

The ecliptic plane is a fundamental concept in understanding the arrangement and movements of planets in the solar system. It is the plane of Earth's orbit around the Sun and serves as a reference plane for the orbits of other planets.

Definition and Characteristics

The ecliptic plane is defined as the imaginary plane that contains the Earth's orbit around the Sun. Since the other planets in the solar system also orbit the Sun, their orbits are inclined at small angles to this plane. The inclination of a planet's orbit is the angle between the plane of its orbit and the ecliptic plane.

1. Inclination of Orbits:

- Most planets in the solar system have orbits that lie close to the ecliptic plane, with slight inclinations. For example, Earth's orbit is defined to have an inclination of 0 degrees, while other planets have small inclinations relative to this plane. Mercury has the largest inclination among the major planets at about 7 degrees, while the other planets have inclinations ranging from 1 to 3 degrees.

2. Ecliptic and Celestial Equator:

- The ecliptic plane intersects the celestial sphere along a great circle called the ecliptic. The celestial equator, another important reference, is the projection of Earth's equator onto the celestial sphere. The angle between the ecliptic and the celestial equator is approximately 23.5 degrees, corresponding to Earth's axial tilt. This tilt is responsible for the changing seasons as Earth orbits the Sun.

Significance of the Ecliptic Plane

The ecliptic plane is significant for several reasons, particularly in the context of planetary movements and alignments.

1. Path of the Sun and Planets:

- The ecliptic represents the apparent path of the Sun across the sky over the course of a year. This path is also the approximate plane in which the other planets move, making it a crucial reference for tracking planetary positions. Because the planets orbit the Sun in nearly the same plane, they appear to move along or near the ecliptic in the sky.

2. Zodiac Constellations:

- The ecliptic passes through a band of twelve constellations known as the zodiac. These constellations form the backdrop against which the Sun, Moon, and planets are observed. The zodiac has historical significance in astrology, where the positions of celestial bodies within these constellations are believed to influence human affairs.

3. Observation and Alignment:

- Understanding the ecliptic plane helps astronomers predict when and where planets will appear in the sky. Because the planets' orbits are close to the ecliptic, alignments of planets (conjunctions) typically occur along this plane. This makes the ecliptic an essential reference for identifying and observing planetary alignments.

4. Seasonal Changes:

- The tilt of Earth's axis relative to the ecliptic plane causes the changing seasons. As Earth orbits the Sun, the angle of sunlight varies, resulting in different lengths of day and night and changes in weather patterns. This axial tilt and the ecliptic plane's relationship are fundamental to understanding seasonal dynamics.

How Alignments Occur

Planetary alignments occur as a result of the intricate interplay between the orbital mechanics of the planets and their positions relative to Earth. While the underlying principles are rooted in the laws of motion and gravitation, the specific conditions that lead to alignments involve several key factors.

Orbital Periods and Synodic Cycles

1. Orbital Periods:

   - Each planet has a specific orbital period, which is the time it takes to complete one full orbit around the Sun. For example, Earth's orbital period is approximately 365.25 days, while Jupiter's orbital period is about 11.86 Earth years. The differing orbital periods mean that planets move through their orbits at different speeds.

2. Synodic Periods:

   - The synodic period of a planet is the time it takes for the planet to return to the same position relative to Earth and the Sun. For instance, the synodic period of Mars is approximately 780 days, meaning that roughly every 780 days, Mars and Earth return to the same relative positions in their orbits.

Alignment Types and Conditions

1. Conjunctions:

   - Conjunctions occur when two planets appear close together in the sky from Earth's perspective. This happens when the planets are on the same side of the

 Sun and aligned along the line of sight from Earth. Conjunctions are classified as inferior or superior, depending on the planets involved. Inferior conjunctions involve an inner planet (Mercury or Venus) passing between Earth and the Sun, while superior conjunctions involve a planet on the far side of the Sun.

2. Oppositions:

   - Oppositions occur when Earth lies directly between the Sun and another planet, typically an outer planet like Mars, Jupiter, or Saturn. During opposition, the planet is fully illuminated by the Sun and appears bright and prominent in the night sky. Oppositions are prime times for observing these planets because they are closest to Earth and fully visible throughout the night.

3. Transits:

- A transit occurs when an inner planet (Mercury or Venus) passes directly between Earth and the Sun, appearing as a small black dot moving across the Sun's disk. Transits are rare events because they require the planet to be in conjunction with the Sun and also aligned with the ecliptic plane. Transits of Mercury occur about 13 times per century, while transits of Venus occur in pairs eight years apart, with more than a century between pairs.

Mathematical and Geometrical Considerations

1. Orbital Inclinations and Nodes:

- The orbits of planets are slightly inclined relative to the ecliptic plane, which affects the likelihood and frequency of alignments. The points where a planet's orbit crosses the ecliptic plane are called nodes. Alignments are more likely to occur when planets are near these nodes. For transits, both Mercury and Venus must be at or near their nodes for the transit to be visible from Earth.

2. Elongation and Apparent Motion:

- The elongation of a planet is the angular distance between the planet and the Sun as seen from Earth. Maximum elongations occur when Mercury or Venus reaches its greatest angular separation from the Sun, providing optimal viewing conditions for these inner planets. The apparent retrograde motion, where a planet appears to move backward in the sky, also influences the timing and visibility of alignments.

Predicting Alignments

1. Astronomical Calculations:

  - Astronomers use precise mathematical models and calculations to predict planetary alignments. These calculations take into account the orbital parameters of the planets, such as their semi-major axes, eccentricities, inclinations, and periods. By modeling the positions and motions of planets, astronomers can determine when alignments will occur and their visibility from different locations on Earth.

2. Software and Tools:

  - Modern technology has made predicting alignments more accessible. Astronomical software and mobile applications allow both professional astronomers and amateur stargazers to track planetary positions and forecast upcoming alignments. These tools provide detailed information about the timing, location, and viewing conditions for conjunctions, oppositions, and transits.

3. Observational Opportunities:

  - Alignments provide unique observational opportunities. For example, during a conjunction of bright planets, such as Venus and Jupiter, the planets can be seen close together in the twilight sky, creating a striking visual spectacle. Oppositions offer the best conditions for observing outer planets with telescopes, revealing details of their atmospheres, moons, and rings.

Chapter 3

The June 3, 2024 Planetary Parade

Overview of the Event

The planetary parade on June 3, 2024, is a rare celestial spectacle where six planets of our solar system—Mercury, Venus, Mars, Jupiter, Saturn, and Uranus—align in a remarkable display visible from Earth. Such alignments, where multiple planets appear in close proximity in the sky, are relatively uncommon and offer skywatchers an extraordinary opportunity to witness the grandeur of our cosmic neighborhood.

Rarity and Significance

The alignment of six planets is a noteworthy event due to its rarity and the visual impact it creates. While individual planetary conjunctions are more common, occurrences where multiple planets align simultaneously are infrequent and highly anticipated by astronomers and stargazers alike. The last time six planets aligned in this manner was over a decade ago, making the June 3, 2024, parade a unique and memorable event.

Profiles of the Six Planets Involved

Each of the six planets participating in the planetary parade possesses unique characteristics that contribute to the overall spectacle of the event. Understanding these planets and their positions relative to Earth enhances the viewing experience and appreciation of the celestial alignment.

1. Mercury

- **Position:** Closest to the Sun and usually challenging to observe due to its proximity to the Sun's glare.

- **Appearance:** Appears as a small, bright dot near the horizon shortly before sunrise or after sunset.

- **Orbital Characteristics:** Orbits the Sun quickly, completing an orbit approximately every 88 Earth days.

- **Observing Tips:** Look for Mercury low in the eastern sky just before dawn or low in the western sky after sunset. Use binoculars or a telescope to enhance visibility.

2. Venus

- **Position:** Second planet from the Sun and often referred to as the "evening star" or "morning star" due to its bright appearance.

- **Appearance:** Appears as a brilliant point of light, outshining all other celestial objects except the Sun and the Moon.

- **Orbital Characteristics:** Orbits the Sun in approximately 225 Earth days, moving more slowly than Mercury.

- **Observing Tips:** Look for Venus in the western sky after sunset or the eastern sky before dawn. Its luminous glow makes it easily visible even in urban areas.

3. Mars

- **Position:** Fourth planet from the Sun and known for its distinctive reddish hue.

- **Appearance:** Appears as a bright, reddish-orange object in the night sky.

- **Orbital Characteristics:** Orbits the Sun in approximately 687 Earth days, making it slower-moving compared to Earth and Venus.

- **Observing Tips:** Look for Mars in the southeastern sky in the evening hours. Its reddish coloration helps distinguish it from nearby stars.

4. Jupiter

- **Position:** Fifth planet from the Sun and the largest planet in our solar system.

- **Appearance:** Appears as a bright, non-twinkling point of light, often outshining all other celestial objects except Venus.

- **Orbital Characteristics:** Orbits the Sun in approximately 12 Earth years, moving slowly across the night sky.

- Observing Tips: Look for Jupiter in the southern sky in the late evening or early morning hours. Its four largest moons, known as the Galilean moons, may be visible with binoculars or a small telescope.

5. Saturn

- **Position:** Sixth planet from the Sun and known for its distinctive rings.

- **Appearance:** Appears as a moderately bright point of light, distinguishable by its rings when viewed through a telescope.

- **Orbital Characteristics:** Orbits the Sun in approximately 29 Earth years, making it one of the slowest-moving planets visible from Earth.

- **Observing Tips:** Look for Saturn in the southeastern sky in the late evening or early morning hours. Its rings are best observed using a telescope with moderate magnification.

6. Uranus

- **Position:** Seventh planet from the Sun and characterized by its bluish-green coloration.

- **Appearance:** Appears as a faint, blue-green point of light, often requiring binoculars or a telescope to distinguish from surrounding stars.

- **Orbital Characteristics:** Orbits the Sun in approximately 84 Earth years, making it a challenging target for casual observers.

- **Observing Tips:** Look for Uranus in the eastern sky in the late evening or early morning hours. Its faintness necessitates dark, clear skies for optimal viewing.

Viewing Tips and Equipment

Observing the planetary parade requires careful planning and consideration of various factors to maximize the viewing experience. Whether using the naked eye, binoculars, or a telescope, employing the right techniques and equipment enhances the clarity and enjoyment of this rare celestial event.

1. Timing and Location

- ***Early Morning or Late Evening:*** Planetary alignments are best observed in the early morning hours before sunrise or the late evening hours after sunset. Check local sunrise and sunset times to determine the optimal viewing window.

- ***Clear Horizon:*** Choose a viewing location with an unobstructed view of the horizon, preferably away from tall buildings, trees, or mountains, to observe planets near the horizon during twilight.

2. Observation Techniques

- ***Naked Eye Observation:*** Start by identifying the brightest planets, such as Venus and Jupiter, without optical aids. Scan the sky for other planets using their distinct colors and relative brightness.

- **Binoculars:** Binoculars provide enhanced magnification and can reveal details such as planetary phases and moons. Use a stable support, such as a tripod, to steady the view and prevent hand tremors.

- **Telescopes:** Telescopes offer detailed views of planets and their features, including cloud bands on Jupiter and the rings of Saturn. Experiment with different eyepieces and filters to optimize the view for each planet.

3. Sky Conditions and Light Pollution

- **Dark Skies:** Choose a viewing location away from urban light pollution to maximize visibility, particularly for fainter planets like Uranus. National parks, observatories, and rural areas offer excellent dark-sky opportunities.

- **Weather Forecast:** Monitor weather conditions and cloud cover forecasts to ensure clear skies during the viewing period. Clouds can obstruct the view and diminish the visibility of celestial objects.

4. Safety Precautions

- **Solar Viewing Safety:** Exercise caution when observing planets near the Sun, such as Mercury and Venus, to avoid accidental solar viewing. Never look directly at the Sun without proper solar filters or observing techniques to prevent eye damage.

- **Equipment Handling:** Handle telescopes and binoculars with care to avoid accidental damage or injury, especially when adjusting mounts or changing eyepieces in low-light conditions.

5. Patience and Enjoyment

- **Take Your Time:** Planetary alignments unfold gradually over time, so allow ample time for observation and enjoyment. Bring snacks, beverages, and comfortable seating to enhance the viewing experience during extended sessions.

- **Share the Experience:** Invite friends, family, or fellow astronomy enthusiasts to join in the viewing experience and share insights and observations. Consider hosting a viewing party or attending a local astronomy event to connect with the community.

Chapter 4

Observing the Alignment

Best Times and Locations

Observing the planetary alignment requires careful planning to ensure optimal viewing conditions. The best times to observe the alignment depend on the specific location and local sky conditions. Here's a guide to identifying the best times and locations for observing this celestial event:

1. Timing:

- *Early Morning Hours:* For observers in the Northern Hemisphere, the best viewing times typically occur in the early morning hours before sunrise. Look for the planets low in the eastern sky during twilight.

- *Late Evening Hours:* Observers in the Southern Hemisphere can enjoy the alignment in the late evening hours after sunset. Planets will be visible in the western sky during twilight.

2. Location:

- **Dark Sky Sites:** Choose a viewing location away from urban light pollution to maximize visibility. National parks, rural areas, and observatories offer excellent dark-sky opportunities for observing celestial events.

- **Clear Horizon:** Select a location with an unobstructed view of the horizon to observe planets near the horizon during twilight. Avoid tall buildings, trees, or mountains that may obstruct the view.

Necessary Equipment: Binoculars, Telescopes, Apps

While the planetary alignment may be visible to the naked eye, using binoculars or telescopes enhances the viewing experience and allows observers to see more details. Additionally, several smartphone apps can assist in identifying and tracking celestial objects. Here's a breakdown of the necessary equipment for observing the alignment:

1. Binoculars:

- *Magnification:* Choose binoculars with moderate magnification (7x to 10x) to enhance the visibility of planets and their moons.

- *Stability:* Use a tripod or stable mount to steady the view and prevent hand tremors, especially when observing planets near the horizon.

- *Field of View:* Select binoculars with a wide field of view to capture more of the night sky and easily locate celestial objects.

2. Telescopes:

- ***Aperture:*** Opt for a telescope with a larger aperture to gather more light and provide clearer views of planets and their features.

- ***Magnification:*** Experiment with different eyepieces and magnification settings to achieve optimal views of planets, moons, and planetary features.

- ***Tracking Mount:*** Consider using a telescope with a motorized tracking mount to automatically follow the motion of celestial objects across the sky, minimizing the need for manual adjustments.

3. Smartphone Apps:

- ***Sky Maps:*** Install astronomy apps such as SkySafari, Stellarium, or Star Walk to access interactive sky maps that identify planets, stars, and constellations in real-time.

- ***Planetarium Mode:*** Use the planetarium mode feature to simulate the night sky based on your location, date, and time, helping you locate and track celestial objects during the alignment.

Challenges and Tips for Observation

While observing the planetary alignment can be a rewarding experience, several challenges may arise, including weather conditions, light pollution, and equipment limitations. Here are some tips for overcoming common challenges and maximizing the success of your observation:

1. Weather Conditions:

- ***Check Weather Forecasts:*** Monitor weather forecasts and cloud cover predictions to ensure clear skies during the viewing period. Plan observation sessions on nights with favorable weather conditions to maximize visibility.

2. Light Pollution:

- ***Choose Dark Sky Sites:*** Select viewing locations away from urban areas and light pollution sources to improve visibility, particularly for fainter planets like Uranus.

- ***Use Filters:*** Consider using light pollution filters or nebula filters on telescopes to minimize the impact of artificial light and enhance contrast when observing celestial objects from urban areas.

3. Equipment Limitations:

- ***Practice Observing Techniques:*** Familiarize yourself with your binoculars or telescope and practice observing techniques before the alignment event. Experiment with different magnifications, eyepieces, and focusing methods to achieve optimal views.

- ***Maintain Equipment:*** Ensure that your binoculars or telescope are clean, properly collimated, and aligned before each observing session to maximize image quality and clarity.

4. Patience and Persistence:

- ***Be Patient:*** Planetary alignments unfold gradually over time, so be patient and allow ample time for observation. Take breaks, relax, and enjoy the experience of exploring the night sky.

- ***Stay Persistent:*** If weather conditions or other factors prevent observation on the scheduled date, don't be discouraged. Planetary alignments occur periodically, providing multiple opportunities for observation in the future.

Chapter 5

The Role of Each Planet in the Alignment

As the planets of our solar system converge in a rare alignment, each celestial body plays a unique role in shaping the cosmic spectacle. From the elusive inner planet Mercury to the distant ice giants Uranus and Neptune, each planetary participant contributes its distinct characteristics to the celestial choreography. In this chapter, we delve into the individual roles of Mercury, Mars, Jupiter, Saturn, Uranus, and Neptune in the grand alignment of June 3, 2024.

Mercury: The Elusive Inner Planet

Mercury, the closest planet to the Sun, often remains elusive to observers due to its proximity to the solar glare. However, during the June 3 alignment, Mercury emerges as a prominent participant, offering a fleeting glimpse of its swift orbit around the Sun. As the innermost planet,

Mercury completes an orbit approximately every 88 Earth days, tracing a rapid path through the sky.

Despite its small size and proximity to the Sun, Mercury exhibits distinctive phases similar to the Moon, ranging from a thin crescent to a nearly full disk when viewed from Earth. During the alignment, Mercury's position relative to the other planets provides a striking visual contrast, highlighting its role as the messenger of the gods in ancient mythology.

Mars: The Red Planet

Mars, often referred to as the "Red Planet" due to its rusty hue, commands attention as one of the most captivating celestial bodies in the night sky. During the planetary alignment, Mars takes center stage, showcasing its distinctive color and prominent features, including its polar ice caps, dusty plains, and towering volcanoes.

As the fourth planet from the Sun, Mars occupies a crucial position in the alignment, offering a striking contrast to the brighter inner planets and the gas giants beyond. Observers can marvel at

Mars' distinctive reddish glow as it stands out against the backdrop of the starry heavens, symbolizing humanity's enduring fascination with the mysteries of the cosmos.

Jupiter: The Giant Planet

Jupiter, the largest planet in our solar system, exerts a powerful influence on the alignment with its massive size and gravitational pull. As the fifth planet from the Sun, Jupiter commands attention with its immense gaseous atmosphere, swirling storms, and iconic Great Red Spot.

During the alignment, Jupiter serves as a celestial beacon, guiding observers' gaze towards the outer reaches of the solar system. Its four largest moons—Io, Europa, Ganymede, and Callisto—add to the spectacle, appearing as tiny points of light against the backdrop of the gas giant's immense disk.

Saturn: The Ringed Planet

Saturn, renowned for its breathtaking ring system, captivates observers with its exquisite beauty and celestial elegance. As the sixth planet from the Sun, Saturn graces the alignment with its dazzling rings, composed of icy particles and rocky debris that orbit the planet in a delicate dance.

During the alignment, Saturn's rings present a mesmerizing sight, casting a slender shadow on the planet's golden-hued surface. Observers can marvel at the intricate patterns and subtle hues of the rings as they catch the sunlight, adding a touch of celestial splendor to the cosmic tableau.

Uranus and Neptune: The Distant Ice Giants

Uranus and Neptune, the distant ice giants of the outer solar system, contribute their enigmatic presence to the planetary alignment. Located far beyond the gas giants Jupiter and Saturn, Uranus and Neptune embody the outermost reaches of our celestial neighborhood, shrouded in mystery and intrigue.

During the alignment, Uranus and Neptune may appear as faint points of light, challenging observers to discern their subtle hues of blue and green against the backdrop of the starry sky. Despite their remote location, Uranus and Neptune remind us of the vastness and diversity of the solar system, inviting contemplation of the unknown realms that lie beyond.

Chapter 6

Future Planetary Alignments

The allure of planetary alignments extends beyond the present moment, captivating astronomers and enthusiasts alike with the promise of future celestial spectacles. In this chapter, we explore the upcoming planetary alignments in 2024 and 2025, examine rare multi-planet alignments on the horizon, and delve into long-term predictions and historical patterns that shape our understanding of cosmic phenomena.

Upcoming Alignments in 2024 and 2025

In the near future, celestial events promise to dazzle observers with their celestial splendor. In 2024, sky gazers can anticipate a rare planetary parade, as six of the solar system's planets—Mercury, Venus, Mars, Jupiter, Saturn, and Neptune—align in a stunning display of cosmic harmony. This convergence offers a unique opportunity to witness the intricate dance of the planets as they journey through the vastness of space.

Following closely on the heels of the 2024 alignment, 2025 brings another celestial treat with a series of planetary conjunctions and oppositions. From the close pairing of Venus and Mars to the dramatic opposition of Jupiter and Saturn, the celestial calendar promises a wealth of astronomical delights for observers to explore and enjoy.

Rare Multi-Planet Alignments in the Future

While individual planetary alignments are relatively common, rare multi-planet alignments offer a truly extraordinary celestial spectacle. These cosmic gatherings occur when three or more planets align in close proximity, creating a breathtaking tableau of celestial beauty and harmony.

One such rare alignment is the grand conjunction of Jupiter and Saturn, which occurs approximately once every 20 years. During these celestial events, Jupiter and Saturn—the two largest planets in the solar system—appear to converge in the sky, creating a stunning visual display that has inspired wonder and awe for centuries.

Looking ahead, astronomers anticipate a rare triple conjunction involving Mercury, Venus, and Mars in the year 2030. This celestial dance promises to captivate observers with its intricate choreography and celestial elegance, offering a glimpse into the dynamic interplay of gravitational forces that shape our solar system.

Long-term Predictions and Historical Patterns

Beyond the immediate future, astronomers employ sophisticated mathematical models and computational simulations to predict long-term planetary alignments and explore historical patterns in celestial phenomena. By studying the orbital dynamics of the solar system, scientists can forecast future alignments with remarkable precision, providing valuable insights into the cyclical nature of cosmic events.

Historically, astronomers have observed recurring patterns in planetary alignments, such as the Saros cycle—a period of approximately 18 years and 11 days that governs the recurrence of eclipses and other celestial events. By analyzing historical records and astronomical data, researchers can identify these recurring patterns and use them to predict future alignments with confidence and accuracy.

Chapter 7

Historical and Cultural Perspectives

From the dawn of civilization to the present day, planetary alignments have captured the imagination of humanity, inspiring awe, wonder, and reverence across cultures and epochs. In this chapter, we embark on a journey through time to explore the rich tapestry of historical and cultural perspectives on planetary alignments, from ancient observations and myths to their enduring significance in modern-day society.

Ancient Observations and Myths

The fascination with planetary alignments dates back to antiquity, where early civilizations gazed upon the night sky with a mixture of curiosity and reverence. In ancient Mesopotamia, the Babylonians meticulously recorded celestial phenomena, including planetary alignments, as omens of divine favor or impending doom. Their cuneiform tablets bear witness to the alignment of planets and stars, which they interpreted as messages from the gods and celestial signs of earthly events.

Similarly, the ancient Egyptians regarded the movements of the planets as reflections of the gods' will, with each celestial body embodying a divine principle or cosmic force. The alignment of planets held deep symbolic significance in Egyptian mythology, shaping religious rituals, royal decrees, and agricultural practices.

In ancient Greece, philosophers and astronomers such as Pythagoras and Aristotle pondered the nature of planetary motion and speculated on the significance of celestial alignments. Their observations laid the groundwork for the development of Western astronomy, inspiring generations of scholars to unravel the mysteries of the cosmos.

Cultural Significance Across Civilizations

Across the globe, diverse cultures and civilizations have imbued planetary alignments with unique cultural meanings and interpretations. In ancient China, celestial events such as planetary conjunctions were viewed as harbingers of dynastic change and political upheaval, with imperial astrologers meticulously recording and interpreting these cosmic omens.

In Mesoamerica, the Maya developed sophisticated astronomical calendars to track the movements of the planets and predict celestial events with remarkable precision. Their intricate calendars, carved in stone and codified in hieroglyphic texts, reveal a deep understanding of planetary cycles and alignments, which guided religious ceremonies, agricultural practices, and societal rituals.

In India, the Vedic tradition associates planetary alignments with the concept of yugas, or cosmic ages, in which the alignment of celestial bodies influences the spiritual evolution of humanity. Ancient Vedic texts such as the Rigveda and the Puranas contain references to planetary conjunctions and their astrological significance, shaping the beliefs and practices of Hindu culture for millennia.

Modern-Day Interest and Media Coverage

In the modern era, planetary alignments continue to captivate the public imagination, fueled by advances in astronomy, space exploration, and media coverage. Amateur astronomers and professional stargazers alike eagerly anticipate rare celestial events, such as the alignment of planets or the transit of Venus across the sun, which garner widespread attention and media coverage.

In popular culture, planetary alignments frequently feature in science fiction literature, films, and television shows, where they serve as dramatic plot devices or cosmic phenomena of apocalyptic proportions. From the prophetic visions of Nostradamus to the fictional worlds of Star Wars and Doctor Who, planetary alignments loom large in the collective imagination, reflecting humanity's enduring fascination with the mysteries of the cosmos.

Chapter 8

The Impact of Planetary Alignments on Earth

Debunking Myths and Misconceptions

One prevalent misconception is that planetary alignments can trigger catastrophic events such as earthquakes, tsunamis, or volcanic eruptions. However, scientific evidence does not support this notion. Despite the gravitational forces at play during alignments, they are relatively weak compared to other factors influencing Earth's geophysical processes. Studies have consistently failed to find any correlation between planetary alignments and natural disasters, debunking the myth of their destructive potential.

Similarly, some people believe that planetary alignments can influence human behavior and societal events, often drawing connections to astrology. Astrology posits that the positions of celestial bodies at the time of one's birth can influence personality traits and life events. While astrology remains a popular belief system for many, it lacks empirical evidence to support its claims. Scientific research has repeatedly shown no correlation between planetary positions and human behavior, emphasizing the importance of critical thinking when interpreting celestial phenomena.

The Influence on Tides, Seasons, and Astrology

Although planetary alignments do not directly cause natural disasters or influence human behavior, they can have subtle effects on Earth's environment and astronomical phenomena. One notable influence is on tidal patterns, which are primarily driven by the gravitational pull of the Moon and, to a lesser extent, the Sun. While planetary alignments can contribute to variations in gravitational forces, their impact on tides is minimal compared to the dominant forces exerted by the Moon and Sun.

Additionally, planetary alignments can affect the Earth's axial tilt and orbital parameters, influencing seasonal variations in sunlight and climate patterns. However, these effects are gradual and occur over long timescales, rather than as sudden changes triggered by specific alignment events.

Astrology, on the other hand, attributes significance to planetary alignments in shaping individual destinies and global events. According to astrological beliefs, the positions of planets at the time of a person's birth or during significant events can determine personality traits, relationships, and societal trends. However, astrology lacks scientific validity and is considered a pseudoscience by the scientific community.

Scientific Studies and Findings

Despite the lack of empirical support for the astrological claims regarding planetary alignments, scientists continue to study these celestial events for their astronomical significance. Observing planetary alignments provides valuable opportunities to study the dynamics of the solar system, planetary orbits, and gravitational interactions.

Modern astronomical techniques, such as computer simulations and space-based observations, allow scientists to accurately predict and analyze planetary alignments with unprecedented precision. By studying the gravitational effects of planetary alignments, researchers gain insights into the underlying mechanisms governing celestial motion and the stability of the solar system.

Chapter 9

Tools for Stargazers

In the pursuit of exploring the wonders of the night sky, stargazers rely on an array of tools and technologies to enhance their celestial observations. From state-of-the-art telescopes to mobile apps and star charts, these tools serve as indispensable companions for amateur and seasoned astronomers alike. In this chapter, we delve into the world of stargazing equipment, exploring the latest advancements in technology and offering practical advice for aspiring stargazers.

Apps and Software for Tracking Celestial Events

In the digital age, astronomy enthusiasts have access to a wealth of mobile apps and software designed to facilitate celestial observations and stargazing sessions. These apps harness the power of GPS technology and astronomical databases to provide real-time information on celestial events, planetary positions, and upcoming phenomena.

One such app **is SkySafari,** a comprehensive planetarium app available for both iOS and Android devices. SkySafari offers a vast catalog of stars, planets, and deep-sky objects, allowing users to simulate the night sky in stunning detail. With features such as time-lapse animations, augmented reality overlays, and telescope control capabilities, SkySafari is a versatile tool for stargazers of all levels.

Another popular app is Star Walk, which offers an intuitive interface and immersive stargazing experience. Star Walk uses augmented reality technology to overlay constellations, planets, and celestial objects onto the user's device screen, providing a captivating glimpse of the night sky from any location on Earth.

For those interested in astrophotography, apps like **PhotoPills** offer advanced planning tools and calculators to help photographers capture stunning images of the night sky. With features such as moon phase calendars, star trail simulations, and exposure calculators, PhotoPills is a must-have app for photographers seeking to immortalize the beauty of the cosmos.

Using Star Maps and Charts

One of the most widely used star atlases is the Sky & Telescope Pocket Sky Atlas, which features detailed maps of the entire night sky, including thousands of stars, galaxies, and nebulae. Compact and portable, the Pocket Sky Atlas is an indispensable companion for stargazers venturing into the great outdoors.

For astronomers interested in specific celestial phenomena, specialized star charts such as the Messier catalog offer detailed information on prominent deep-sky objects, including galaxies, star clusters, and nebulae discovered by the French astronomer Charles Messier in the 18th century. These charts provide invaluable guidance for locating and identifying celestial objects during stargazing sessions.

www.ingramcontent.com/pod-product-compliance
Lightning Source LLC
Chambersburg PA
CBHW050237230526
45470CB00005B/2000